TLC

Teams, Leaders, and Change

Dr. Amanda H. Goodson
Dr. Yvette Rice
Odetta Scott, MBA, MSOD

AMANDA GOODSON

TLC: Teams, Leaders, and Change
by Dr. Amanda H. Goodson, Dr. Yvette Rice, and Odetta Scott, MBA, MSOD

Edited by Adam Colwell's WriteWorks, LLC: Adam Colwell and Ginger Colwell
Cover design by Decatur Printing Solutions
Typesetting by Inktobook.com
Published by Amanda Goodson Global

Printed in the United States of America
Hardcover ISBN: 978-1-951501-14-3
Paperback ISBN: 978-1-951501-15-0
eBook ISBN: 978-1-951501-16-7

While the authors have made every effort to provide accurate internet addresses at the time of publication, neither the publisher nor the authors assume any responsibility for errors or for changes that occur after publication. Further, the publisher does not have any control over and does not assume any responsibility for author or third-party websites or their content.

What people are saying about TLC: Teams, Leaders, and Change

Throughout my career, I have had to navigate being one of the only women in engineering, let alone one of the very few women in executive engineering leadership. One thing I have learned in my career is that finding your own internal voice, and being authentic to your own unique and powerful style, allows you and your team to excel.

This book helps recognize where misnomers, systematic bias, and your own self-talk may limit women from reaching their full potential. Applying the tools and guidance contained within this book will help you learn your own authentic style of technical leadership, embrace it, and build upon it. Let your voice be heard! You are more powerful and have more influence to create the career and work culture that you want.

Rebecca Stoner

Sr. Director of Engineering, Collins Aerospace Interiors

***

Dr. Amanda H. Goodson, Dr. Yvette Rice, and Odetta Scott are perfect examples of people who are passionate about Science, Technology, Engineering, and Mathematics (STEM) education and assuring that all children have the opportunity to share in these exciting fields of studies and career opportunities. Their efforts to encourage young women (especially women of color) is extraordinary. Having professionally known Amanda for many years, I still remember when she and I were among the few people of color in Space Shuttle launch readiness reviews at the Kennedy Space Center. Even then, I could see that she was destined for greatness.

I fully endorse Amanda, Yvette, and Odetta in their efforts to "burn the light brighter" for women in technology.

James L. Jennings

Former Associate Administrator, Institutions and Management, National Aeronautics and Space Administration (NASA) Headquarters

***

Even after the Women's Suffrage and Women's Rights movements of the twentieth century, the twenty-first century still finds women lagging behind their male counterparts in moving into leadership positions, especially women of color. In *TLC: Teams, Leaders, and Change,* Dr. Amanda H. Goodson, Dr. Yvette Rice, and Odetta Scott instantly recognized that something needed to be "demystified" within women to offset the inner conflict preventing them from pursuing advancement opportunities.

TLC: Teams, Leaders, and Change is one of the best mentoring/self-help/leadership books I have read that is focused on the professional development of women. This book is an easy read with well-structured chapters written to present seven vital principles with supporting developmental steps, motivating stories, and personal examples. The principles are born out of over 60 years of collective cooperate experience, personal challenges, private struggles, and public triumphs.

This book simplistically and systematically presents practical, progressive insights to ensure success, and the authors engage readers with excitement and passion. With succinct focus, *TLC: Teams, Leaders, and Change* conveys how women in technology can tap into the greatness inside of them with a little TLC (tender loving care). It is a must read for any woman who knows greatness is within her but needs a roadmap to help guide her onto that path of greatness.

Dr. Claudette C. Owens

President/CEO, Quantus Link, LLC

Table of Contents

Acknowledgments

Dr. Amanda H. Goodson

I dedicate this book to all the people that have supported me along the way. I especially want to thank my family, particularly my husband, Lonnie, and son, Jelonni, who have been so patient with me, supported me, and loved me through so many experiences. I appreciate them for trusting me, being faithful to me, and loyal to the purpose that we are intended to pursue together. I'd like to thank my co-laborers of WeTECH Rocks for working diligently to get this product over the line. For those people who encourage me, equip me, excel me, and propel me, I am eternally grateful to you! To my mother, sister, and late father, I love you. To my Heavenly Father, I will continue to do what you request of me. I trust that you were pleased. Also, to the many TLC readers, enjoy life and accelerate!

Dr. Yvette Rice

I give God all of the glory and honor, for without Him, I could do nothing. A loving thanks to Amanda and Odetta for pursuing the vision of WeTECH Rocks, and to Adam Colwell and his wife, Ginger, for jumping on board to get this book across the finish line. Thank you to all of the women who so boldly pursue their dreams in Science, Technology, Engineering, and Mathematics (STEM) industries. A special thank you to my husband, Bishop Sam Rice, for encouraging me to spread my wings and soar to my God-given destiny. Also, to our children Sharné, Samuel Christopher, and Amber, thank you for your loving support.

Odetta Scott

I dedicate this book to my village—my spouse, my family, friends, mentors, sponsors, and coaches. Thank you so much for pouring into me and helping me push and extend beyond where I could even see. Lord, through this work and I all I do, I give you the glory, honor, and praise. To my WeTECH Rocks co-founders, I extend my gratitude to both of you for being willing vessels and accepting the call for this consortium and in creating this program. I look forward to continuing to blaze a path with you as we accelerate women in Science, Technology, Engineering, and Mathematics (STEM)!

Foreword

On July 29, 2020, four leaders from the world's big technology companies appeared before the United States Congress in an anti-trust hearing meant to question alleged anti-competitive behavior and unhealthy market domination. It became an often robust exchange that exposed the runaway nature of global capitalism as well as the tolerance society has had toward untransformed leadership and lack of modeling. The hearing revealed that bullying has become corporate culture.

For many, what stood out most about the hearings was that the quartet—Facebook Chief

Executive and Founder Mark Zuckerberg, Google Chief Executive Sundar Pichai, Apple Chief Executive Tim Cook, and Amazon Founder and Chief Executive George Bezos—were men being quizzed by a Congressional committee also made up predominantly of men. Just a few years after pressure was exerted by the #MeToo movement for corporate culture to embrace inclusivity, what worried me about the hearing was the exclusive focus on the companies' alleged corrosive market practices—not about how the gender makeup of companies in the United States (and, by extension, the world) has birthed the corrupt leadership behavior that was being questioned. Men and women bring very different ethos to leadership.

What needs to be of concern to us in corporate leadership is whether when we talk about a Boy's Club, do we equally see a need for a Girl's Club? Boys have linear and non-linear clubs going by names such as "drinking buddy," "acquaintance," "fellow team supporter," "fellow grunge fan," and/ or "we met on a plane." What do we have? We are often expected to discover ourselves and our business acumen within societies that don't tolerate women who down tequila shooters after a networking cocktail party. She has to be home and read the kids a bedside story while the husband folds his sleeves, loosens his tie, and drinks to a stupor.

We are oriented by a society where our so-called acquaintances must espouse some of the values we were burdened with when we were born. "Is she married?" "What are you doing with an unmarried woman while you have a family?" Men don't get asked these questions, and, in those rare cases when they are, it's not by their parents, but intimate partners. Women are not expected to love sports, not even the Women's National Basketball Association or the Ladies Professional Golf Association. However, it's considered good for women to be courtside cheerleaders or to strut into a boxing ring to announce the next round.

As a result, nobody is talking to young girls. They spend their days with their faces illuminated by cellphone screens consuming junk that does not amount to role modeling or mentorship. All it does is filibuster their independent thought. It's a space we need to claim. More than a decade ago, a South African mobile network provider company, Cell C, conducted a campaign called, "Take a Girl-child To Work," an annual day where workers were encouraged to bring their daughters to work. As years dragged on, sulking emerged that, while the girl-child was learning the ropes, who was mentoring the boy-child? It was a legitimate concern that should have resulted in an effort to locate both the boy-child and the girl-child in the matrix of proper leadership parenting. We didn't need to weaken the right to strengthen the left, which is not what that campaign was doing in the first place.

Mentorship is the best shot we have at dragging

everyone to a space of self-affirmation where they are allowed to unleash their potential, fall, and get picked up. It is through such interventions that we can change the complexion not only of technological fields, but shift the whole paradigm toward a society receptive of women leadership—which will mean seeing company boards trustful of female chief executives and women leaders keen on lifting up their own and changing corporate values.

As we strive for inclusivity, and for women to take their space in Science, Technology, Engineering, and Mathematics (STEM), we need sound, compassionate, and committed leadership who can encourage and create a conducive environment for women to have a voice and dignity in the technical field marketplace. Dr. Goodson, Dr. Rice, and Mrs. Scott's *TLC: Teams, Leaders, and Change* is a worthwhile response to this challenge.

In the wake of comprehensive advances in the fields of science, technology, engineering, and mathematics, the world has changed at an alarming rate. Beyond raising families, women are taking up crucial roles in corporate and business arenas. Opportunities are available today for women that have never existed before. As they multiply, we can grab them with the knowledge that we can and are able to do so.

Allow me to close with my philosophy: "I know who I am. I know who I am not. I can never forget where I come from. I know where I am going. To reach for my dreams and conquer the challenges, I will dare to be different, and, with proper guidance, direction, and God on my side, nothing is impossible. I am indeed a leader of future leaders."

Sibongile Rejoyce Sambo
Founder and CEO, SRS Aviation Ltd.

It is time!

*I*t all started when we were together in Washington D.C. at the BEYA STEM Global Competitive Conference in February 2020.

Hosted by *Black Engineer* magazine, the Council of Historically Black Colleges and Universities Engineering Deans, Lockheed Martin Corporation, and Aerotek, the conference was designed to create connections between students, educators, and professionals. It also facilitated partnerships with individuals and their local STEM (science, technology, engineering, and mathematics) resources.

The Black Engineer of the Year Awards (BEYA) were also presented as part of the three-day event.

This annual conference has attracted top professionals and students from all over the nation and every field of science, engineering, and technology for more than two decades. During the 2020 event, we made some discoveries that changed the trajectory of our lives—especially when it comes to what we want to do for girls and women in technology.

———

An interview by researcher Joanna Barsh, the director emerita and senior partner from McKinsey & Company, revealed data analytics gathered from individuals nationwide showing that a distinct gap still exists between women moving into leadership roles in technical fields as compared to men. This gap was particularly pronounced with African American women.

As engineers who have worked, or are still working, in technical jobs in government, industry, nonprofit, and academia settings who have traveled globally as women in technology, we knew that had to change.

We also discovered that only seven percent of women who graduated from college in 2016 earned a STEM degree, while STEM degrees remain more popular with men in general. Currently, the National Science Foundation reports that women comprise 43 percent of the United States workforce for scientists and engineers under 75 years of age, and 56 percent for those 29 or younger. But when it comes to the women's presentation skills, business acumen, leadership of themselves and teams, and executive presence, they still lack confidence to go after certain leadership positions when compared to men.

We instantly recognized that something needed to be demystified within women to offset the inner conflict preventing them from pursuing advancement opportunities.

Finally, of the 14 engineering schools of higher learning that brought teams to compete in the Advancing Minorities' Interest in Engineering (AMIE) Design Challenge, we found that less than 20 percent of those participating were young women, even though those females brought the same technical competence to the event. Again, we were told the percentage was so low because the other women didn't feel they could compete. Even more, their aspirations were not to become a CEO, a technical leader, or a senior fellow. They perceived themselves as only being capable of lesser options.

We saw that our schools are filled with young women who don't see themselves as technically competent and able to hold their own in technical fields. That, of course, is a misnomer.

———

Astonished, we thought of those African American female leaders who have gone before us: Linda

Gooden, who has been black engineer of the year and is the retired executive vice president of Lockheed Martin's Information Systems & Global Solutions business area after nearly 40 years working in the aerospace and defense industry; Stephanie C. Hill, a recent black engineer of the year and current senior vice president of Enterprise Business Transformation at Lockheed Martin; and Dr. Nandi Leslie, a senior principal engineer at Raytheon, serving as a researcher in the Network Science Division at the U.S. Army Research Laboratory whose Ph.D. is in applied and computational mathematics.

We thought of ourselves and what we've been able to achieve through our shared love for leadership and personal development, and our searing passion for young girls and women to be able to soar to a place that creates an advantage for them in technology.

We asked ourselves, "What can we do together to become a global leader in bridging the gap for the advancement of women in technology?"

———

That's when WeTECH Rocks, a Women in Technology consortium, was born—with the mission of accelerating value delivery, skills development, and strategy execution for leaders and entrepreneurs in STEM. Its goal is to teach women and young girls who are emerging in business, industry, the community, and in academia how to be better leaders.

TLC: Teams, Leaders, and Change is the first book release to champion that goal.

We want to be a catalyst for women in technical careers, and we want to support women in such a way that their confidence is built with pointers to help them be stronger in their respective fields. We also desire to do the following:

- Develop best practices and solutions for women to get better jobs and achieve better placement in STEM in the areas they desire.
- Create opportunities where senior women leaders in the community can come and share ideas and wisdom through mentorship, advocacy, and coaching.
- Facilitate a transfer of knowledge and skills development to help young women be competent and confident.
- Make competition irrelevant because these women will bring something to the table that is so vital, it will be amazing.

———

We believe it is time for women to not just stand up, but to stand out and be heard. We are not done yet. We are not there yet. It is imperative for women to look at what they can do, as opposed to what they have been *told* they can do, and then take their place to do mighty things as professionals.

It is time for us as women to have a *voice* and *dignity* in the technical field marketplace through how we communicate, how we share, and how we lead to shift workplace culture.

It is time to bring hope and encouragement for those already in technical fields to burn the light even *brighter* for women who aspire to what they have already achieved.

In this book, we will deliver seven principles that will help you become a catalyst to the TLC (teams, leaders, and change) you will *perpetuate* as a woman in technology—while giving yourself the TLC (tender loving care) you need to *go forward* as a confident, capable woman in technology.

———

1. **Identify your style, exploit your strengths, and recognize your differences to manage them well.** How you show up daily and interact with people will provide direction to how you execute, inspire, and strategize in a business setting. Are you strong? Are you communicative? Are you detailed? We will help you understand your realized and unrealized strengths and characteristics, how you feel, think, and behave, and how you produce actions and behaviors from others.

---

2. **Acquire cultural awareness and clarity to become more effective as a technical professional.** Most work today is performed by teams. We don't look or perform the same, and research shows that diverse teams outperform non-diverse teams by up to 30 percent. Diversity appears in all types, including experiences, thoughts, gender, age, and culture. In order to motivate these diverse teams successfully, those who perform on them have to be able to connect. Having cultural awareness provides a way to connect with your team to perform better and to navigate and deconstruct conflict.

---

3. **Communicate and connect early and often by speaking up and speaking out in professional meetings and by using technical writing to create excellence and advance your career.** Communication is how we verbally or nonverbally pass information from one person to another, and how well you communicate will affect your ability to lead. In the marketplace, communication skills are critical to add to or change the culture in an organization. Your behavior, speech, appearance, and body language are

keys to delivering executive presence in a way that says, "I belong here." The power of words, the phrases that you use, and the things you should and shouldn't say to bring agreement to what's going on in the room can affect change and influence how you are treated and respected.

———

4. **Strive to be great!** Never give up. Never give out. Never give in. Greatness is inside of you regardless of whether it has been highlighted. Being great allows you to stop thinking the same thing over and over again. Then, when fear shows up, you've got to disable it, disarm it, and kill it so that you can be the person you are destined to be. Your knowledge is needed— and the greatness in you requires you to exude confidence as much as you can. You owe it to yourself and to all women in technology.

———

5. **Use strategic planning, which is foundational and critical to map your future.** Strategic planning is about imagining the big picture to see your desired end state and having big and bold ideas as a leader, or in your team, that map into the organization's current and future objectives. As you learn to solve a problem

in a way it has never been solved or refine an existing technology or idea, you will make the competition irrelevant and move forward.

———

6. **Invest in technical and professional development, knowing that continuous improvement is the key to success.** When you think about a financial investment, you want to yield a return of some kind. Likewise, as you invest in yourself technically and professionally, you should see a return on that investment. You never want to become stagnant or stale. With technology changing as quickly as it is, you must be able to understand and translate those changes. That becomes the rationale, or driver, behind the phrase, "What's in it for me?" to make sure you are always growing and developing. There is always something new to learn, and feedback is to be seen as a gift that gives you an opportunity to take it to heart and improve in a way that will set you apart from your peers.

———

7. **Receive ongoing technical coaching to give you best practices, and mentors to help you navigate to places you never thought you'd go.** Your technical acumen has to be

sharp. It has to be something that adds value to the organization. A coach can look at the whole landscape, understand the practices that have been done before, and show you how you can apply your technical ability and prowess so that you can distinguish yourself and be sought out for your genius, proclivity, and organization. A mentor can assist and help you to do things beyond your original thoughts so that you can get to the front of the line faster, head and shoulders above your counterparts.

———

It is time for girls and women in technology to be lifted to a place where they can think higher, better, and broader, planting themselves in a totally different place than where they are now.

This is your time to jump in! It is our privilege and joy to get you there.

Find out exactly why WeTECH Rocks! Let's get started.

Dr. Amanda H. Goodson
Dr. Yvette Rice
Odetta Scott, M.B.A., MSOD

1

Identify your style, exploit your strengths, and recognize your differences to manage them well.

*L*eaders may be bold, or they may be deep and analytical. There are other leaders who are gregarious, animated, and strong at developing relationships—and then there are those who are positive and may have an upbeat mantra for each day: Magnificent Monday, Terrific Tuesday, Wonderful Wednesday, Thankful Thursday, and Fantastic Friday.

What kind of leader are you? Or, are you yet to come to the realization that you are a leader?

Truth is, you don't have to have a title to be a leader. You can lead from any chair! As you are authentic to your true self and learn how to be an influencer, you can get in the room, get to the table, and make an impact. Influencers surround themselves with people as good as, or better than, they are. They are also servants who are willing to learn, listen to people who have gone before them, and stretch themselves.

Many of you were influencers starting at a very young age, you just didn't realize those leadership tendencies were already developing within you. Perhaps it was as a child when you influenced other children on which games to play, or maybe it was in middle or high school when you developed tendencies others wanted to emulate or follow. Whatever it was, the ability to influence and lead others was there, and it still is today.

Attributes that can define your leadership style include defined focus areas such as strategy, continuous improvement, and talent development, with each role infused by a passion and energy to make a difference. You may need to adapt and flex those attributes and your style depending on the situation you're in or the problems you're facing at that time. As you do this, you'll discover you possess a wider range of attributes and styles than you initially believed.

There are several assessments (such as DiSC) that can help you discover those attributes, define them, and begin to put them into use. The assessments can also help you navigate through how to work with other people.

Working with other people not only raises your awareness about them, but it also heightens your knowledge about yourself. It helps identify your triggers and how you manage your emotions and the emotions of others with whom you interact. The capacity to be aware of, control, and express your emotions in interpersonal relationships with others is called "emotional intelligence"—and it is vital to effective leadership.

Daniel Goleman is an expert on emotional intelligence as well as an internationally known psychologist and lecturer. He says there are six styles of leadership. "Coercive leaders demand immediate compliance. Authoritative leaders mobilize people toward a vision. Affiliative leaders create emotional bonds and harmony. Democratic leaders build consensus through participation. Pacesetting leaders expect excellence and self-direction. And coaching leaders develop people for the future."[1] Meanwhile, the four attributes of emotional intelligence are self-awareness, self-management, social and situational awareness, and relationship management. The ability to understand yourself and your emotional intelligence as it pertains to your leadership attributes and style will help you flex when you need to adjust in order to get others galvanized and engaged.

To know how to best identify your style, exploit your strengths, and recognize your differences, we recommend you conduct a **SPOT analysis** of yourself. In advance of doing the analysis, ask yourself:

- "What are the things I am doing today that I like the most?" followed by,
- "What are the things I least like to do?"

The first question allows you to list your "productive ways of being" to identify how you act and what you produce that adds value to others. The second question causes you to ascertain your "non-value-added ways of being" by honestly evaluating areas where your attitude or actions do not add value to those around you.

Perhaps you tend to get angry when you feel like your back is against the wall. Maybe you become controlling when your sense of security is threatened. These ways of being almost always stem from an area where your belief system has been challenged or violated.

In addition, ask yourself,

- "If I were to go to my stakeholders—those who know me well such as my spouse, parents, siblings, classmates, and co-workers—what will they say I do well or don't do well?"

Hearing what others have noticed about you all your life can often reveal things you've never recognized in yourself.

Next, the SPOT analysis takes what you learned about your "productive ways of being" and "non-value-added ways of being" to look at yourself from the standpoint of your strengths (S), potential areas for improvement (P), opportunities (O), and threats (T). Take a sheet of paper and write a big plus sign (+) on it, then place an S, P, O, and T in each of the different quarters.

Now, fill in each quadrant using the following guidelines:

- Strengths represent areas where you believe you are most effective and therefore place you in a position of advantage. These identify things you can *leverage*.
- Potential areas for improvement can come from places where you feel you are at a disadvantage and, therefore, need more training or development. You can take a class or read a book and *learn* how to address these areas.
- Opportunities constitute areas that you believe you can exploit to even greater advantage. *Listen* to others and strive to apply and make the most of what they tell you.
- Threats are elements, usually external, that can damage or even endanger your ability to fulfill your mission. As you recognize these, your job is to *lessen* them.

As you look at your leadership style and the benefits of organization, collaboration is key. You were likely a collaborator from when you were young and playing with your friends and family members, in high school with clubs and organizations, or even in the classroom. Each served as a collaboration lab in which you were exposed to different things to help develop and strengthen the person you were always meant to be. As you come into the professional world, you have already learned the flexibility, adaptability, relatability, and likeability factors you bring to the table.

When you pull together a team as a leader, working together with them to leverage their differences and exploit their strengths is very important. Once you understand their differences and strengths, you create a capability for yourself, your team, and your organization that will propel it forward in ways you never thought possible.

When you have an employee that is new to the workplace, or when your team doesn't yet have the skills they need or are a little laid back, your style of leadership has to be different with them than it is with someone with 20 years of experience who knows what they are doing. You'll need to be more prescriptive and directive with the newer team member. On the other hand, you'll more easily delegate trust to the veteran employee. Even though you are still in the leadership role, you'll say, "Okay, I'm going to give you the resources and the access you need, and I am going to pull back and let you work."

In the area of collaboration, you need to have the ability to give your team a voice in the organization. That means you shouldn't always be the one with all of the answers or the only one with thoughts and ideas. Establish opportunities for your team to collaborate so each one can bring their best to the project or the problem you are tasked to solve.

There will be times as a leader when you feel hesitant about saying, "I don't know," or "I want help." You may think of it as a sign of weakness if you have to ask someone else for answers. It is not. Don't allow yourself to be put in a spot where you feel like you can't reach out to someone to gain knowledge or perspective. Instead, strive to be a self-aware leader who knows where you may not be as strong and surround yourself with people who are stronger in those areas than you. That will accelerate the team to move forward because every base is covered, and you will end up receiving the best of the best in every strength to be more efficient, effective, and high performing as a team.

That said, in times of crisis, you have to be confident enough, both in yourself as a leader and in the information your team has given you, to bring an end to the collaboration. That is when your authoritative leadership style steps in to say, "We no longer have time to get everyone's opinion. I trust and place value on what you have given me, but the time has come where I have to make a deci-sion." It all comes back to having an ability to shift your leadership style when necessary. You look at the information you have, ascertain the risk, and decide where to go. Your emotional intelligence in crisis situations as a leader is vital.

A word from Yvette

A great way to mitigate conflict is to employ my five functions of a team required for conflict management. Just use a little GRACE:

G = Good vibes and trust. You are there for each other and those you serve.

R = Recognition of each team member that produces a healthy respect for other's opinions different from your own.

A = Accountability and maintaining a sense of responsibility. You must do your part without judging your teammates.

C = Character. Your character exemplifies who you are as you fulfill your purpose on the earth.

E = Excellence. Your work must always be excellent.

Finally, what do you do if your team has already been chosen for you? It's already in place and, frankly, is not the team you feel you need to accomplish your task. That's when you revector them, putting them in different seats as you help them navigate to find the role that is right for them and best for the assignment. There have also been occasions as a team member when one of us realized we were not the right fit for the vision of the leader. That's when it's time to revector and ask, "How can I help you to be successful?" That often means being moved to a different role as a result.

———

While we'll look at connecting in more detail a bit later in *TLC: Teams, Leaders, and Change*, there is a way you should connect as a leader when you are looking at your strengths and identifying the differences that will resolve problems and challenges in the organization. In this case, think of connecting like building a house. Start with the foundation and the framing, and when you know they are right, you can then go from there to create the rest of the building. This requires flexibility, relatability, and emotional intelligence. In addition, you'll want to connect often with your team to make sure you're in alignment and that everyone has what they need to do their job. If you have a team member who is not as direct or up front as others, connect more frequently with that individual so that everything keeps going in the right direction.

In his book, *The First 90 Days: Critical Success Strategies for New Leaders at All Levels*, Michael Watkins recommends promoting yourself as you connect. Tell yourself, "I am getting ready to go into this new environment. I am going to emotionally connect and put myself in that place." As you promote yourself, you must see conflict as being productive. There are different levels and types of conflict. If it is chaotic, for example, then your job is to stay above it. That will help you put things in the right perspective and then be able to create a strategy and goals to put those chaotic things in the right buckets and the right places. You'll be able to compartmentalize it so that it makes sense.

As you look at ways to mitigate conflict by using your strengths and exploiting your differences, always strive to be part of the solution. View conflict as an opportunity for creativity, innovation, and more productivity. Let your self-talk rise above the challenges you see. In addition, asking questions to get a better understanding of the conflict allows you to get down off your ladder emotionally so that you don't escalate issues when things are hot and heavy around you. When dealing with conflict, there are differences in opinions, values, and skills that pull in strong emotions and trigger behaviors. Therefore, if you don't deal with conflict well, you may need to have a conflict analysis done to find out how you can better handle conflict and identify ways you can improve.

NOTES

1   Daniel Goleman, "Leadership That Gets Results," (Harvard Business Review Classics). Harvard Business School Publishing Corporation, 2017.

2

Acquire cultural awareness and clarity to become more effective as a technical professional.

Technical professionals encounter diversity in various skills, experiences, and exposure. Therefore, you must develop your strategy and professional acumen according to what is happening within the culture you are in. This requires you to have cultural awareness and cultural clarity to exercise cultural leadership.

Cultural awareness starts by recognizing that not all cultural differences are visible. You can observe practices and behaviors, but other aspects such as beliefs, education, ideas, and perceptions are not as easily observed. Think of it as an iceberg. There's much floating beneath the surface that simply isn't seen, but that doesn't make it any less important. As you recognize surface culture and subsurface culture, you can begin to demystify some of the more challenging aspects of your particular culture so that you can better understand your team and how to maximize your role in that team.

Cultural clarity is also vital because today's workplaces operate from a place of cultural inclusion as well as diversity. Diversity appears in many places, including experiences, thoughts, gender, age, and culture. You need to understand the framework of the culture, as well as discern where it is going in the future, to be able to operate effectively in that culture and leverage your experience as a technical professional. Having cultural clarity provides a way to connect with your team to perform better and to navigate and deconstruct conflict.

————

Therefore, WeTECH recommends that you complete a diversity profile to look at the top 10 things that are important to you and to the organization you work for (or are interested in being employed by) from a cultural standpoint. It's essential to take the time to identify your cultural behaviors and practices. Then take a coworker, or someone on your team, and identify their practices and behaviors. Write them down, analyze them, and then see what behaviors need to grow or pivot to get the best from you and your team. It'll help you understand the strengths of each person on your leadership team as well as those on your support team. As much as possible, you need to have conversations about diversity and talk about it openly.

A diversity profile will also cause you to identify what you bring as a professional that is of high value and medium value and discern where you fit. Then you can see how that comes together, connects, and causes a cadence so the organization performs well as you get to know your team, build those relationships, and recognize that everyone has something to contribute that adds value.

The Harvard Business Review, in its March 2018 issue, described a great culture as one that has alignment between behavior, systems, and practices, where the entire organization walks the talk from the senior leadership down. Businesses that are high performers have figured out how to drive efficiency by operating from a cultural perspective as opposed to saying one thing and doing another. They make sure that the tension between behavior systems and practices is tight. There should not be a lot of slack. When everybody knows what the organization is expecting at the board level, the business level, and the individual level, they can act and perform accordingly.

————

It's no accident that gender diverse companies are more likely to outperform non-gender diverse companies by 15 percent, while ethnically diverse companies are more likely to outperform less ethnically diverse companies by 35 percent.[1] Cultural awareness and diversity are not necessarily synonymous with each other, but they are connected and important to be able to add value and impact change. We've heard it said on numerous occasions that "culture eats strategy for breakfast," an idea first coined by Peter Drucker, the founder of modern management. A business strategy is connected to a company's vision, which is really important. The vision informs employees and others where the company is going. If culture can indeed eat strategy, it goes to show how important culture is to an organization.

————

In addition, Glassdoor, a website where individuals seek employment and current and former employees anonymously review companies, states that 67 percent of job seekers believe diversity is a vital factor when evaluating companies and job offers.[2] It's not unusual for prospective employees, when examining a company, to look at the organizational chart and then shy away when they look at the leaders of that company and see no diversity in the executive staff. However, consider that you might become the person

A word from Amanda

During my six years as Director of Safety and Mission Assurance at the National Aeronautics and Space Administration (NASA), not only was I the first African American woman to ever hold that position, I also received some significant and humbling honors, including an Exceptional Service Commendation and Federal Employee Woman of the Year recognition. I also spoke often, traveling to places such as Georgia and Mississippi and sharing on subjects like personal warranty (doing what you say you are going to do) and technical careers in engineering, all serving to inform the public about the value and impact of the space shuttle program.

But perhaps my biggest achievement was the way I made a difference in the lives of those I interacted with by influencing the establishment of a more accessible culture at NASA. People said I was down-to-earth and easy to talk to, which I felt allowed me to demystify communication with NASA leadership. That created an opportunity for me to understand different cultures and how my culture may be seen by others. With that, there was the need for adaptability and a level of understanding of how others think and perceive what I think. It was rewarding work.

who breaks that barrier. There may be a purpose for you to be there. Instead of drawing back in fear of being overlooked or undervalued, it could be an opportunity for you to break that glass.

When United Technology Aerospace Systems merged in 2018 with Rockwell Collins, the president of the newly formed organization shared an organizational chart that was not ethnically or gender diverse with the entire population at senior leadership levels. This chart was shared with over 70,000 employees. It wasn't gender or ethnically diverse. When asked if the company was pursuing the Paradigm for Parity® movement initiative, he said it was, and he reaffirmed the importance of gender equality in leadership ranks.

Paradigm for Parity is a coalition of business leaders dedicated to addressing the corporate leadership gender gap. The coalition is made up of CEOs, senior executives, founders, board members, and business academics who are committed to achieving a new norm in the corporate world: one in which women and men have equal power, status, and opportunity.

Its ultimate goal is to achieve full gender parity by 2030, and there's powerful evidence that this will have a positive impact beyond culture. Research shows that companies with both women and men in senior leadership positions have superior financial and stock market results.

A word from Odetta

Being a woman of color in engineering, a male-dominated industry, I have found that I don't have to be liked, but I do want respect for the value that I bring to my work and to the companies I serve. The skills that I learned to go in, make an impact, be respected, and still influence the process and drive it to the outcome needed have been very eye opening for me, and I have gained more culture awareness from these experiences.

That makes it paramount for me to honor the concept of "each one reach one" in my position as a role model for cultural leadership. I try to be an example by going out and providing exposure in the community. I introduce young women to the engineering field and recognize other women who are already making significant impact in STEM.

As women, we add incredible value which may be dissimilar to that of men, but our value should not be minimized or marginalized. In fact, it should be elevated and fully appreciated for what it is. When you bring your authentic self and your capabilities to the table, you can integrate your different ideas into a solution, and that is what drives innovation.

As a new or potential leader, remember that you can add value even being the newest member of the team. Be aware of who you are. I've also come to see that it's not as much about the direct value you bring, but about the greater impact you can have by developing others. You are only one person, but if you reach out and develop many other people, the impact and value that has on the culture can be maximized and optimized. To help others, you have to be able to ask questions that will cause them to identify "aha" moments and thoughts to uncover ideas and desires they may not be aware of yet. Seek to listen to and understand what the individual wants to do, and then ask questions to help them get there.

NOTES

1   McKinsey & Company, "Why Diversity Matters," Vivian
    Hunt, Dennis Layton, and Sara Prince, January 2015.
2   Glassdoor Inc. blog, November 17, 2014.

3

Communicate and connect early and often by speaking up and speaking out in professional meetings and by using technical writing to create excellence and advance your career.

ommunication happens in two ways: verbal and nonverbal. Verbal communication includes your tone of voice and the words that you speak. Nonverbal communication involves body language, hand motions, eye contact, and facial expressions, all of which serve to modify what you are communicating. In fact, listeners will not only look at what we say, but how we say it. Often, we are not aware how much non-verbal communication informs how credible we come across and whether or not what we say really matters to those who are listening.

That established, the person you communicate with the most is yourself. Whether you realize it or not, you can paint a defeated picture of yourself in the marketplace when you tell yourself you're not good enough or smart enough—especially as a woman.

Katty Kay and Claire Shipman expose the gender confidence gap in their book, *The Confidence Code*. Their research reveals that even women at the top of their professions still question their worth, saying they were "just lucky" or that they somehow "slid through" or feel like a "fraud." Furthermore, they state that, while women tend to underestimate their abilities, men overestimate theirs by 30 percent. Our communication, and how we first speak to ourselves, can be detrimental to our ability to succeed.

How do you defeat those negative voices? You have to learn to develop the right self-talk. Before you can communicate with anyone else, you have to learn to communicate in a positive way with yourself. Former First Lady of the United States, Eleanor Roosevelt, has been quoted as saying, "No one can make you feel inferior without your consent." As women, we cannot give consent to that negative voice that reinforces our insecurity or inferiority. Why? There will be times others will not see the greatness in you; therefore, you must learn to encourage yourself through positive confession. Putting those confessions in front of you on the wall or on a mirror while you are getting dressed in the morning allows you to speak positive things to yourself.

———

How well you communicate will also affect your ability to lead. For example, communicating an organization's vision in a succinct way is critical. If we cannot express ourselves in a way that generates passion within our teams relative to the vision, we'll upset the direction of the organization.

Ask yourself:

- "Am I creating energy?"
- "Am I getting them excited and igniting a desire within them to understand why they are a part of the team and the organization, and that what they bring to it really matters?"

In addition, your communication skills add to or change the culture in an organization. Through WeTECH, we have a training packet available called The EDEN Culture. This culture is a place where your vision for the organization is communicated, and your team members' purposes are well defined and declared. It shows you how to consistently use the Language of LUV (Listening, Uniting, and Valuing)™ to promote personal and professional growth among leadership and peers in the organization. It causes you to consider:

- "Does my communication promote that I value the organization and value the peers and the leaders of the team?"
- "Am I listening, or am I just waiting for an opportunity to speak without really hearing what the team member or peer is trying to say?"
- "Does my communication and body language show that I value the organization and the members by promoting personal growth and leadership?"
- "Have I communicated that the members of my team can come and talk to me openly at any time and that I will listen without judgment?"
- "Do I clearly communicate gratitude and appreciation?"

––––––

In their book *Winning*, Jack and Suzy Welch express the importance of voice, dignity, and candor. They say people want the opportunity to speak their minds and have their ideas, opinions, and feelings heard, regardless of their nationality, gender, age, or culture. That goes back to diversity. People also want to be respected for their work, effort,

A word from Amanda

When I was at NASA, I worked with a brilliant engineer from Germany named Bob. He designed rocket engines, and I noticed him reading constantly from three-by-five index cards he carried with him in his shirt pocket whenever there was a lull in the meeting.

"What do you have on those cards?" I asked.

He held a few of them out to me. I saw the small but legible handwriting in ink. "These are mathematical formulas," he said with his decidedly staccato accent. "Engine designs like this used to be drawn up on napkins or with paper and pencil. I read these again and again so that I know them."

I took a lesson from Bob, got a package of my own 3 x 5 cards, and started carrying them with me. I didn't write equations on mine, though. Instead, whenever I heard a leader say something influential or use a power word or phrase, I immediately wrote it down. Later, I played games with myself using the cards, putting tick marks on each one every time I used what was written on it, or using them like flash cards to retain and memorize a term like "off nominal condition." I created ways to make that content my own, transpose it to increase my knowledge, and improve how I communicated and connected with others while enhancing my executive presence.

and individuality as you communicate with them as a leader. Candor, meanwhile, involves honesty and openness as it relates to communicating the performance of your team members. If you don't communicate candidly, you are doing a disservice to them because they need to know the areas where they can improve.

———

When it comes to connecting, leadership expert John C. Maxwell, in his book *Everyone Communicates, Few Connect*, states that the ability to connect with others goes beyond words. It expresses if a leader cares about their team, is willing to help their team grow professionally, and if the leader is trustworthy. If you are not showing these attributes, a wall goes up and your communication fails. Not surprisingly, Jay Hall, Ph.D. of the Leadership Management Institute and founder of Teleometrics, conducted a study of 16,000 executives to demonstrate the correlation between a leader's success and their ability to connect well with others. His findings showed that a leader must connect with people in order to influence them.

Finally, your behavior, speech, appearance, and body language are keys to delivering an executive presence that says, "I am supposed to be here, and I have value to add to this organization." According to Dr. Carol Kinsey Goman, Ph.D., an international keynote speaker and author of *The Silent Language of Leaders*, executive presence for women requires an added dimension that can include something as basic as how you sit at a conference table. She said most women sit around a conference table with legs crossed, elbows into their waist, hands together on their lap, and their shoulders slightly rounded. Women condense their bodies. She believes such body language can infer signs of insecurity and a lack of confidence. Yet Marie Forleo, a motivational speaker, author, and web television host, encourages women to "show up in every single moment like you're meant to be there."[1]

A former participant at one of our workshops was on the leadership team of a company. Whenever she came into a conference room to sit down, she brought her notebooks and all the things she would need and put her items on the table just like everyone else. Once, a gentleman sitting next to her said, "You're taking all of the space at the table." He wasn't used to a woman finding her place so comfortably. Her response was, "I'm supposed to be here just like you."

When a woman sits at a conference table and there are a lot of people in the room, we have often found that the guys will come and try to crowd her out. Therefore, figure out in advance what your real estate is at the table, and when you arrive, put your phone on one side, your agenda on the other, and your laptop in the middle. Make sure you have all the territory you're supposed to have. As you do this, you'll also communicate to those around you, "I may not know as much as you do about something right now, but I am going to bring something to the table one day." You are preparing and positioning yourself for that time.

Don't forget: the reason you are in the room is because someone decided you were smart and had something to offer. If it is nothing but a penny, and everyone else has 99 cents, your penny is going to make their dollar.

————

A couple more points to remember about communication is that when you're using social media or email, you can leverage these same communication principles to guard and enhance how you interact with others and portray yourself and your organization. Something as simple as using correct spelling, capitalizations, and punctuation in emails or texts can make a huge difference in how others perceive you. You don't want to send a message with errors so embarrassing that you feel compelled to correct and resend it. All caps in either one of those communication methods, for example, can communicate yelling when you didn't intend to do so. In addition, more than one exclamation point at the end of a sentence is not viewed as professional writing and can mistakenly communicate anger or frustration.

Second, words have power. The phrases you use don't go away. If someone has been hurt or offended by your words, they may read or rehearse those words over and over again. In addition, when you write, be sure to use words appropriate to your readership. If your writing is too technical for a layperson audience, your communication will go over their heads and you will not be understood. Plus, if you've been asked to submit a two-page memo on an issue, meet that format precisely. It will teach you how to write succinctly and with utmost clarity.

We've had people say to us, "You write well for an engineer." They explain, "Most scientists and engineers write equations and numbers, but when it comes to focusing on getting a point across, or explaining something in writing, they have a problem communicating or connecting." Writing well will set you apart. Use software to make sure your content is grammatically correct and communicates exactly what you want it to say. Wait a couple of minutes before you send anything, giving yourself an opportunity to step away and then come back and reread it. Sometimes we write when we're emotional or excited. Other times, you'll notice something you misinterpreted.

Amanda always reminds us to employ emotional intelligence skills as a leader to Stop, Think, and Act as a writer.

- Stop and assume a positive intent.
- Think about it from the standpoint of how both of you can best connect.
- Act on your discernment.

You can apply the same GRACE standards Yvette shared in Chapter 1 to your writing. Finally, we recommend taking a communication class or workshop to improve your abilities. WeTECH offers these opportunities.

NOTES

1   https://iamfearlesssoul.com/20-marie-forleo-quotes/

4

Strive to be great!

*I*t's already been mentioned in *TLC: Teams, Leaders, and Change,* but it can't be repeated enough: you bring something to the table. You have a body of knowledge that other people don't. As you innovate and create who you are and leverage who you are, realize that you are *needed*. Please do not make light of that fact.

The greatness in you requires that you exude confidence in yourself and what you offer. There is a difference between being assertive and being aggressive. Assertiveness is strength under control. Aggressiveness is strength without control. Let's says there's a woman in a room of engineers having a technical conversation. She possesses more knowledge in a certain area than the others, yet one of her colleagues keeps challenging her. A woman with strength under control will listen to everyone's feedback to the degree to which it is appropriate and then thank them for their comments. Then she can communicate her area of expertise in a way that is articulate without being aggressive. She will do so with a cool, calm, confident tone of voice.

However, a woman who has strength without control will react. "What do you mean by that? I am the one that knows this. I am the expert!" Her voice gets louder, her inflection is raised, and she tries to shut everyone down. Later, those in the meeting will think she was angry, but she wasn't. She was passionate, but that passion had driven her to the point where she was not in control. The constant challenges to her knowledge and authority were the trigger.

An aggressive woman can be perceived as attacking, threatening, and disrespectful. An assertive woman stands up for herself or her views by being controlled and respectful, effectively connecting and communicating in a way where her self-value is equal to how she values others.

———

To achieve greatness, you need to be in control. In addition, you need to be confident in yourself and in your abilities. In the book, *Lean In: Women, Work, and the Will to Lead*, by Nell Scovell and Sheryl Sandberg, the authors equate going to the top as a leader with playing on a jungle gym. You are to be fearless about getting on that jungle gym because you could be the greatest person that adds value in your organization. Author, speaker, and ministry leader Joyce Meyer declared, "Do it afraid."[1]

To be confident and to strive to greatness, you also have to deal with and address impostors or impediments that you may face. There may be times when you hear people speak eloquently or fluently on a topic and feel you are at the bottom of the heap in comparison. That's when you come against that feeling and speak something that will add value. You may not give exactly what they give, but you will give the best of what you have at that moment.

There are also false narratives that can continually play in your mind. "You can't do it." "You are not smart." "You are not good at it."

To be great, you have to shut that conversation down.

That was then. This is now.

Live in the now—not in the past

Create a new narrative for yourself.

When Amanda was younger, she was told by a school counselor that women were not supposed to be engineers and that she wouldn't make it because she was a woman. She shut that down with two words: "Watch me." She refused to receive that conversation and let it fester in her heart. Instead,

A word from Yvette

John Joseph IV, executive director of the Decatur-Morgan County Entrepreneurial Center in Decatur, Alabama, has been such an example to me of a great leader, motivator, cheerleader, coach, and mentor. It was John, along with the good Lord, who exhorted me, at 60 years of age, to get a master's degree in business administration through the Jack Welch Management Institute at Strayer University. My background was varied with a predominance in technology, and even though I had a doctorate's degree and a master's degree in theology, it was not enough. I felt like I needed additional leadership and business credentials. I was looking to grow my leadership acumen to be able to offer greater diversity to the customers I expected to serve.

When I was up at one, two, three o'clock in the morning, I reminded myself why I wanted to be better at what I am doing. I called it delayed gratification—giving up some things in the present to accomplish greater things later. The professors I had at the Jack Welch Management Institute greatly influenced me, seeing and pulling out the best in me, whether it was through a challenging assignment or by asking questions to help me develop my critical thinking process. Now, I

> am teaching and sharing leadership qualities
> with others—and I am indeed better at what
> I am doing. That's greatness.

she knew what she had to do. It was something she
wanted to do. She felt strongly about it—and she
became great.

To further develop the greatness you possess, be
sure to look at your rituals and disciplines. Every
day, you should be disciplined to operate in your
place of greatness. Some people choose to become
expert in systems. For others, it may be avionics or
navigation and control.

What is your discipline—and how can you build
rituals into your life that foster that discipline? Author,
public speaker, life coach, and philanthropist Tony
Robbins speaks of CANI, or "Constant and Never-
ending Improvement." When you wake up every
day, choose one thing that you are going to do better.
Place a reminder on your phone or your calendar or
use a picture that you look at every day to represent
that one thing. If you want to be an expert at a particular discipline, you must develop a constant and
never-ending improvement in your ritual, working
on it daily until it becomes part of your fiber.

To create a discipline that'll take you to greatness, ask yourself Amanda's six "P's."

- Passion: Can I love it? Can it give me lift?
- People: Will people around me notice it, ask
  about it, and even pay for it?

- Preparation: Am I willing to get the education, experience, and exposure to develop and maximize it?
- Position: Will I put time and effort in to set goals, and then deconstruct those goals into objectives, to achieve it?
- Purpose: Am I driven toward that place, knowing why it is important and certain that my heart knows it is right for me?
- Proclivity: Am I bent or inclined to lean that way?

―――――

Doing your SPOT analysis will help you see your discipline. Discovering your FRESH WILL (see next chapter) will help you identify what you need to work on to apply CANI to your actions each day. It is also true that developing the ability to reframe something you may feel was not as successful as you had hoped will bring continuous improvement.

When Odetta mentored Jean, a female executive in the male-dominated aerospace and defense industry, Jean was dealing with a common struggle for a female leader: a tendency to feel like she had to be perfect and even overcompensate to be respected and heard. In her case, and similar ones we have encountered through WeTECH, reframing (the ability to take a scenario or a situation and express it differently) is key. Odetta told Jean, "You are enough, but what I need you to do is get out of your own way so you can be better positioned to think about the things that make you different and add value, and then speak from those things." Understanding and declaring what differentiates you is a leadership characteristic that works regardless of gender. "You just have to go in and be you," I say. "If you continue to be you, they will see that value."

It took about three months or so of repeatedly beating that drum to set up that cadence in her life, but now Jean feels it, believes it, and acts on it. She is a great leader and is very humble. Though it has taken some time, Jean is now doing well in reframing her situation. Instead of focusing on what she doesn't bring and what she can't do, she is focusing on what she can bring and do. She is focusing on her own greatness.

———

The other thing to keep in mind is that as you step toward something, you are also moving away from something else. Therefore, if you are moving toward something that does not make you great and is not part of the narrative you have created for yourself, then you are going to end up in a place you don't expect to be.

Leave being stale behind.

Leave being stagnant behind.

Leave things that are negative behind so that you can step into greatness!

A word from Amanda

When looking at personal and professional greatness, I often ask people to write what they feel makes them great on a piece of paper, and then take the top five from that list and ask, "What is my methodology, and how can I add value with these things?" I'll then ask them to evaluate what part of these great things describes them as a leader.

I will then make analogies for them using my experiences in NASA with the space shuttle. For example, the engine basically kept the shuttle running smoothly, so you may be the engine in the organization. Or, you may be like a rocket booster that provides lift. The rocket booster on the shuttle provided 80 percent of the lift at takeoff. Perhaps you're the fuel tank of your organization. The fuel tank on the shuttle took up an Olympic-sized swimming pool worth of fuel every 30 seconds, so you may be the one keeping your team moving forward at a fast pace. Or, maybe you're like the shuttle itself, the bird that keeps the team on its mission.

Whatever it is, identify your greatness, accept it, and then maximize it.

In the book she co-authored with Amanda entitled, *How to Unlock Your Full Potential*, Odetta wrote this about greatness:

I believe we have the ability to do everything to meet our destined purpose in life. We have the ability to do mighty things in our work, home, and community when we put forth the effort. We may go through difficult times, but we should not give up on being our best.

Our inherent capability is just like the story of the transformation process from a caterpillar to a butterfly. I believe we have superior, powerful, and effective ability mapped right inside of us. As we seek our destiny, I believe that we discover our full purpose. As we become like the butterfly, we can soar higher and make things happen around us that we never thought we could.

The caterpillar has such a different perspective on life. It has a limited view and limited self-expectation. The caterpillar cannot think in extremes because its expectations are low, slow, and lacking. Although the caterpillar may not be beautiful initially, the inherent sense and the nature of the caterpillar is necessary for the butterfly to get its wings.

Later, in the cocoon stage, it may seem as if movement has ceased and the growth has been severely hindered. This is a stage of uncertainty and uneasiness. This stage might seem to hinder your ability to grow and use your gifts as you need to. One may feel constricted and limited in their growth potential.

At the butterfly stage, one emerges to a place of beauty and excellence. The ability of the butterfly is no longer hindered but released to pollinate and affect others around it. The butterfly has a different perspective in life. Its perspective is one of higher expectations, security, ease in flight, and excelling.

The butterfly emerges to occupy a space that shows great promise, destiny, and awe. It has been released from its cocoon and is now unlocked and able to reach its full potential by getting up, having a different perspective on life, and taking on the opportunity to soar!

I want to take a pause here to say this—for those who think they cannot be good at doing something, this is for you. Know ahead of time that you are awesome and capable of power, creative potential, and excellent ability. When your thoughts are focused on the right destination, you will achieve greatness!

So, see yourself as being transformed already. If you are not there yet, it will be coming soon. Transformation takes place in your mind first. If you see yourself as transformed, then act that way. Get ready. Set. Now, transform!

NOTES

1   Joyce Meyer, "Do it Afraid! Obeying God in the Face of
     Fear," FaithWords Publishing, 2003.

5

Use strategic planning, which is foundational and critical to map your future.

*D*r. Nandi O. Leslie is a senior principal engineer at a prominent aerospace and defense contractor, serving as a researcher in the Network Science Division at the U.S. Army Research Laboratory. Her research interests are focused on cybersecurity and resilience quantification and assessments, explaining what makes networked devices vulnerable to cyberattacks—and why not? As one of the most noted experts of her kind, she is a great inventor and innovator in the areas of cybersecurity and cyberspace. She helped to pioneer using machine learning and other computational modeling approaches to predict, detect, and/or block malicious or anomalous network traffic.

Nandi has driven cyberspace technology to new heights. She was the recipient of the Outstanding Technical Contribution in Industry Award at the BEYA (Black Engineer of the Year Awards) Gala held at the STEM Global Competitive Conference in February 2020. The award is granted to individuals in the workforce, and they are nominated for these awards by their employers.

You can achieve outcomes that will cause you as a woman in technology to grow, mature, and excel over a long-term period by using strategic planning. An essential method WeTECH recommends to move from your current state to a desired, future state is reverse planning via GAP analysis. A tried and true tool, you can use GAP analysis to determine what steps need to be taken to map your future by doing the following:

1. List characteristic factors (such as competencies and performance levels) of the present situation or "what is."
2. List factors needed to achieve future objectives or "what should be."
3. Highlight the gaps that exist between the two and need to be filled.

———

To better understand reverse planning, think of how a GPS works. When we enter the address of our destination into the system, the GPS works backward to find the most direct route to where we want to go. It starts at the end point and progresses back toward the starting point. In doing so, it analyzes the trip by identifying obstacles along the way such as road construction or closures, estimating how long it'll take to get there depending on how much traffic might exist, and coming up with a street-by-street path we can follow and rely upon to complete the trip successfully. Then, as a result of its reverse planning, we drive onward to move forward toward the destination.

It's true that hindsight is always 20/20. If you have already been there, then you will know the steps to get there. That's one of the things that makes reverse planning so powerful and effective.

We can then use the GPS example to explain another benefit of doing the GAP analysis in reverse. As we're driving, sometimes something will come up that'll force us to deviate from the path the GPS planned out for us. It might be something necessary, such as a quick errand. It could be accidental, such as not being able to merge safely over from the left lane when we needed to turn right, forcing us to backtrack. It may be as simple as a spur-of-the-moment choice to take another road for part of the journey because we liked the view it provided. Whatever it was, the GPS recalibrated to get us back on the correct route and keep us there.

In the same way, you will have some detours along the journey to your desired future. The beauty of reverse planning is that it is flexible. It is revisited and adjusted on a quarterly, monthly, even weekly basis, but always in such a way as to keep you headed in the right direction toward the manifestation of your future reality.

———

With that perspective in mind, think back to the findings of your SPOT analysis from Chapter 1, and then look at all those results in light of Amanda's nine FRESH WILL characteristics of her Goodson 9 Block.

Finances	Relationships	Energy
Spirituality	Health	Work
Innovation	Leisure	Long Life

A word from Odetta

I have used strategic planning to help me navigate the journey of my career and my life. It has proven to be an invaluable tool. In the previous chapter, Amanda mentioned positioning: putting time and effort in to set goals, and then deconstructing those goals into objectives, to achieve them.

In 1987, when I first heard Mae Jemison (the first female African American to go into space) had become one of the 15 candidates NASA selected out of more than 2,000 people to enter training to become an astronaut, I left my home in Vicksburg, Mississippi in eleventh grade to attend a magnet school in Columbus, Mississippi to enhance my education. I mapped out a path of objectives and structured my curriculum to pursue my goal to become an astronaut and follow in Jamison's shoes (or, in this case, space boots).

I attended the U.S. Naval Academy in Annapolis, Maryland (as most astronauts at that time were trained in the U.S. Navy), but my time at the Academy was bittersweet. I successfully completed my freshman year, but toward the completion of the following year, I began to experience some medical issues. By my third hospital stay, I realized I wasn't going into space like Mae did in 1992—but I figured I'd do the next best thing as an engineer by helping send others up into space. I used my

> strategic planning mindset to move on to Texas A&M University to finish college and earn my degree in mechanical engineering technology. Today, I leverage that same strategic planning skill set in my personal financial planning and in my fiduciary responsibilities.

1. Finances
2. Relationships
3. Energy
4. Spirituality
5. Health
6. Work
7. Innovation
8. Leisure
9. Long Life

Notice how the first letters of each characteristic spell out the term FRESH WILL. It's great because your scrutinization of each one creates a fresh, renewed purpose to your life that gives you direction and propels you forward. It takes you from your current reality to your future reality and reignites your passion so that you can make a declaration to live out your FRESH WILL and manifest outcomes that bring your future reality to fruition.

———

As we look at strategy, your strategy should not necessarily compete against your coworker or organization. Your strategy should complement it and others so you can make them irrelevant. The thought processes, actions, and skill set that you develop and deploy should make what they do irrelevant. One way this idea is referred to is via "blue ocean strategy," which is defined on the Blue Ocean website as "the simultaneous pursuit of differentiation and low cost to open up a new market space and create new demand. It is about creating and capturing uncontested market space, thereby making the competition irrelevant."[1]

Think of "market space" as being your position in a company as a woman in technology. As you use strategic planning and the other guidance we've given you in this book, you will be able to differentiate yourself and create a space for yourself in the organization nobody else can fill. By making the competition irrelevant, we don't at all mean demeaning the competition or doing anything that has a negative connotation. We are talking about making them irrelevant because they are quite literally not relevant in comparison to you and what you have to offer.

Another way to look at it is that you are creating a space where you are in your own lane. You're not going down a path somebody else made. Consider a track and field competition. A sprinter runs in her own lane. She is competing alongside others, but she won't cross lanes or get in the way of the other competitors. It's completely up to her to race against the clock, and if she is the fastest, it doesn't matter what anyone else does. She is going to win—and, if she's in a relay race, she will successfully pass the baton to her teammate to continue the race.

A word from Amanda

Working on my doctorate's degree, I wanted to work on service leadership. My professors were concerned the topic wouldn't be as diverse or relevant as needed because there was already so much work done on the body of knowledge about leadership. But I saw there wasn't as much known about *stewardship* in the workplace or as a leader.

Therefore, I wrote my dissertation on leadership, focusing on stewardship and, specifically, on individual responsibility. Leaders, I said, are to be accountable to where they are. They know their gifting and greatness, and they are able to stay there, deepening and strengthening their knowledge to make their team stronger. In terms of making our competition irrelevant, our responsibility as leaders is to take care of our lane. As we are good stewards over that which is given to us, we will be distinct and successful.

As a woman in technology, what can you do to make your competition irrelevant? Here are some ideas:

- See things that need to be done that others cannot do or will not do and then *do* them.
- Single yourself out as the one person or resource who can offer what the team or organization needs the *most*.
- Find the path of least resistance (perhaps a position that is less glamorous) but will still advance your career the way you desire and then *pursue* it.
- Seek mentors and sponsors who are two or three levels *above* where you are at the moment and work with them.
- Become someone who can foster connections between the technical and non-technical members of your organization and then *leverage* the networking opportunities that creates.
- Become active in roles most other people in technology *aren't* interested in and then create highly productive teams in those spaces.
- Develop and use your communication skills to *bridge gaps* that exist in the organization.

Each one of these strategies will open doors to you that may otherwise remain closed, and will give you an opportunity to relate to the head (knowledge as you teach), heart (feelings as you relate), and soul (motivations as you walk along-side) of others.

NOTES

1   © Chan Kim & Renée Mauborgne. All rights reserved. https://www.blueoceanstrategy.com/what-is-blue-ocean-strategy/

6

Invest in technical and professional development, knowing that continuous improvement is the key to success.

major focus of your strategic planning as a woman in technology should be to complement and add value to your organization—and continuous improvement and learning is one of the enablers that allows you to do just that, while setting yourself apart and making your competition irrelevant. You want to make sure you are doing everything in the most

A word from Odetta

The 2016 motion picture Hidden Figures highlighted the experiences of three African American women (Katherine Johnson, Dorothy Vaughan, and Mary Jackson) who were called "human computers" for NASA and used their knowledge of mathematics, engineering, and computer science to contribute to John Glenn's historic 1962 orbit around the Earth. Vaughan, played by Octavia Spencer, is recorded in NASA history as its first African American manager. With no external prodding or encouragement, she taught herself to code the early IBM computer, correcting errors no one else could solve to get the system up and running. She understood the need to be ready and was able to pivot quickly based on her vision of how the IBM computer would be used and the needed resources to program and support the work.

Without Vaughan and her personal decision to invest in her expertise, and the key contributions of Johnson and Jackson, Glenn might've died during reentry and the space program could have been delayed or worse. With the help of these women, NASA ended up doing what many once considered impossible during a time when both racial and gender bias were common. Today, you'll find the Mary W. Jackson NASA Headquarters in Washington D.C. and the Katherine G. Johnson Computational Research Facility in the NASA Langley Research Center in Hampton, Virginia. President Barack Obama also presented Johnson with the Presidential Medal of Freedom on November 24, 2015.

efficient and productive way every single day of the week and on every single component or challenge that you are responsible for.

Continuous improvement and continuous learning occur in collaboration with one another. Continuous improvement speaks to organizations and products, while continuous learning is an individual's response to the modern workplace. With things changing constantly, including technology, staff, and even the direction of companies, we have to be agile as technical professionals. When the COVID-19 pandemic hit in 2020, most everyone was not ready or familiar with how to behave and work in that environment. Those who worked in a corporate or office setting had to adjust and learn to work remotely as a great number of facilities were closed down due to social distancing or other factors. Parents with children in day care or elementary school had to figure out how to work and be productive remotely while assuming teaching and other responsibilities that were required with the children at home.

A word from Amanda

I once heard the story of a woman and her friend who attended a technical conference and saw a well-respected scientist at a booth. The woman knew that scientist had spent 60 years working on solutions to problems in the aerospace and defense industry and was one of the most noted people in his field. As she approached the table with her friend, the woman saw that the scientist was selling his memoir. His story and anecdotes were all assembled not as a special-bound book, but in a big spiral-bound notebook.

They exchanged greetings, then she looked at the notebook. "How much does this cost?"

"I'm selling them for $150."

Without hesitation, the woman got out her card to pay for her copy. As she did, her friend started laughing.

She pulled her friend aside. "What's so funny?"

"You're going to pay that much for that man's memoir? That is asinine."

The woman shook her head. "No, it's not asinine. Think about it. Sixty years of this scientist's knowledge and lessons learned are in that notebook. That's priceless, no matter how its packaged."

As her friend stopped laughing, the woman added, "How can you grow if you don't learn from other people's failures and success?"

Investing in your technical and professional development is much the same as a financial investment. When you put money in a bank and earn x amount of dollars, you don't just go in and pull out that investment, saying to the teller on the way out, "Thank you very much!" You keep investing because you want your money to grow.

In the same way, you invest in your future and your longevity through continuous improvement—and you also set the stage for other young professionals to come who you can equip and support.

Change came fast and had to be dealt with quickly. Many professionals spent up to 80 percent of their day on phone calls or video conferences working with their teams and resolving critical issues. They had to become more familiar with technological tools they may not have used as much, or at all, before COVID-19. These professionals discovered the difference that was made as they connected to the global community in unprecedented ways.

In addition to coping with the unexpected, new knowledge or skills also help us see things in a new light and take the next leap forward in innovation, allowing us to better leverage our resources and those of the organizations we serve. The SPOT analysis and other tools mentioned earlier in this book are valuable resources to shore up and strengthen yourself and your abilities. As a leader, you can also leverage assessments (such as DiSC) that can help you gain a better understanding of your leadership style and that of others so you'll know when to flex to achieve the optimal desired outcome.

In order to be innovative technical leaders, we have to stay up to date. What is happening in our industry? What can we learn from that in order to bring additional impact to our companies? A Pew Research Center survey found that 87 percent of workers think it is essential to develop new skills to keep up with the changing workplace.[1] We need to be continuously learning. We need to be implementing continuous improvements.

There are many ways you can develop yourself and stay up to date. One thing you can do is take a certification course in a subject like change management. Qualitative conversation is another great approach. Let's say you are working with one of your customers, and they are unhappy. You don't quite have a solution right now, but you have already shown them the data. Have a conversation about how you can deliver the information to them in a way that is relatable to their values, then bring in your team, brainstorm solutions, and chronicle those ideas. After that, using best practices and lessons learned, present your findings to the customer. A third idea is to work to understand the other personality types around you so you can better present your data in a way that is relatable and appealing to them. This goes back to gaining and using emotional intelligence. Some personalities want straightforward bullet points; others will respond to a more narrative approach. The point is to strive to stand out and create more value.

———

Here are some resulting principles from continuous improvement and learning in a technical professional career. Developing new skills and knowledge can help you "AIM HIGHER."

- A = Advance: If you are always in a mode of working on yourself and your skill set, you will be ready when the opportunity comes.
- I = Improve: Improve your performance with the application of enhanced learning and skills. Some people say you can listen all day long, but until you can teach it to someone else, you really don't know it.
- M = Marketability: Additional learning and successful application makes you more marketable both within and outside of your

organization. Draw off of your additional and tangential skills and experiences and parlay that for the area you wish to pursue.

- H = High Demand: Exploit your strengths well and you will become a person that is always being requested for opportunities. But be careful, because the better you are, the more people will want you on their team.

- I = Innovative: Continuous learning should help you to be more innovative. When we were children, we were given coloring books and crayons and told to color inside the lines. Innovation says, "I don't want to color inside the lines. I want to add something to the other side of the line that makes the picture more appealing."

- G = Genuine: In all your activities, be your authentic self. How many times have you tried to restructure or change something about yourself, only to become more uncomfortable? You don't want to lose yourself.

- H = Helpful: You have a responsibility to reach back and support someone else. Since you are already in the starting block ready to run, you are positioned to charge forward to help those wanting to rotate to a new position or grow.

- E = Excel: Excellence is your friend. Embrace it! You are a cut above. Let it spill over and affect the lives of other people so you can have a more positive impact.

- R = Ready to Act! Engage with your coworkers and organization to apply what you have learned. When you and your team do their best, the products and services you provide for the people you serve will be the best.

NOTES

1   https://www.pewresearch.org/internet/2017/05/03/
      the-future-of-jobs-and-jobs-training/

7

Receive ongoing technical coaching to give you best practices, and mentors to help you navigate to places you never thought you'd go.

In the book, *The Purpose and Power of Mentorship*, compiled by Amanda and Marvin Carolina Jr., 16 authors (including Odetta) who have given and received mentorship came together to teach and declare the incredible difference coaching and mentorship have made in their lives as well as the lives of others.

In her chapter entitled "Give Back," Amanda

talked about the term "Seven Up," referring to the number of mentors or coaches she recommends anyone should have at any one time. She wrote, "I know, that may sound like a lot. But because I believe mentorship is critical in every area— business, industry, academia, at a not-for-profit organization, even for a volunteer—the amount is more than justified. In fact, the very success of people and teams hinge on mentorship. I can tell a well-organized individual or a well-oiled team by the mentorship they've received."

As a woman in technology, you should have a diverse portfolio of coaches and mentors. As noted in Chapter 5, you want to choose people who are two or three levels *above* where you are as your coaches or mentors. Odetta, for example, identified a mentor three levels above her to challenge and motivate her. This mentor was also able to provide a perspective of a landscape where Odetta was less familiar. The mentor was an executive director who had grown in a similar function, yet different from Odetta's, within the organization over several years. As part of the mentoring process, Odetta had to look at her five-year strategic plan and constantly refresh it to ensure she was challenging herself, was flexible not knowing what vectors from the environment would come into play, and that the plan was aggressive but achievable. Odetta was also encouraged to become very specific with what she wanted to do instead of speaking in generalities.

Finally, Odetta was asked to create advocates who were engaged in conversations she may not have been aware of so that she could be properly positioned for the next opportunity.

In challenging herself, Odetta had to be able to engage in productive and respectful conflict conversations to move the needle while holding others accountable, regardless of their level in the organization. Being flexible means having the ability to accept change, lead change, and influence the behavior of others to gain alignment to the ultimate vision. She discovered that regardless of where someone sits in the organization, she has to have at least one person (but preferably many) advocating on her behalf and leveraging their social capital for her. Odetta was very moved by how transparent her mentor was with her and says that experience still resonates with her and was monumental to her career.

———

If you stretch a rubber band vertically, the top of the rubber band represents where you want to go technically. The bottom of the rubber band is where you are. The difference between the two parts symbolizes that creative tension that causes you to feel pressure. Applied rightly and correctly, that pressure will cause you to propel yourself upward and onward.

Three individuals who can address that tension

A word from Amanda

Former NASA deputy administrator, James Jennings, was associate administrator over all of NASA, and he was also the center director for the Kennedy Space Center where we did all the space shuttle launches, when I was with the organization. I was fortunate and blessed to have James as both a coach and mentor. We didn't have regularly set meetings with one another, but when I did see him, we talked about different challenges I was facing, and he helped me.

I learned three valuable lessons from James. First, he always told me, "Work really hard because your input and value delivery counts. Your expertise counts. People are looking to your leadership to make a difference in their lives." Secondly, he taught me to get the job done right. Even when we are working hard, we must watch to make sure we're not going in the wrong direction. People often use the analogy, "Practice makes perfect," but if we're practicing incorrectly, the outcome will be anything but perfect. Instead, practice makes *permanent*, so we need to get it right. The last thing James taught me is to have fun. Proper leisure and rest create harmony in our lives.

In a January 2019 speech, Jennings emphasized the needs for technology professionals to treat everyone with respect, learn as much as possible about the organization and its people, get to know your management/ teacher and make sure they know you, and expose yourself to as many positive experiences as you can. "The best way to get to a better position is to excel at the one you are currently in," adding that "leadership is the power to evoke the right response in other people."

An African American, James grew up in Alabama and received a bachelor's degree in mathematics and a master's degree in business administration from Alabama A&M University before earning another master's degree in administrative sciences from the University of Alabama.

in a significant way for women in technology are coaches, mentors, and sponsors. We've heard it said that a coach talks to you, a mentor talks with you, and a sponsor talks about you.

Choosing a technical *coach* is important. A coach provides developmental support. Think of a coach on a football field or a basketball court. They are looking at every aspect of the game. They watch what the other players are doing and how they are making their plays. They see the whole field or court. Likewise, a technical coach will see intricacies across a broad spectrum that others don't necessarily notice. They will see how you fit into an organization or how your professional skills can be adapted or augmented, so you can go and help the team perform and create a competitive advantage for your organization.

The tension between the two points in our rubber band example is where the technical coach comes in to give you best practices so you can relieve that pressure through your knowledge, skills, and experiences. The right coach will cause you to be refreshed, renewed, restored, and reinvigorated.

A technical *mentor* can help you navigate to places where you never thought you could go. Think again of the rubber band. You want to aim it toward the target and hit the mark when you let go of it. You don't do this blindly. You want to do it purposefully and effectively, and a mentor enables you to do just that. A mentor may not necessarily be someone you like as a person, but you know they bring something to the table that will help to seed the skill set that you need to grow. It's great to have good relationships with your mentors, but there may be some things about them that may cause you to take pause. That's okay. The focus is learning and allowing them to invest in you. A mentor is someone with a more focused view who thinks about ensuring that you have the right skills and knowledge, and who will teach you if you want to develop a particular area of expertise. The mentor talks with you from a more in-depth perspective.

A word from Yvette

Beginning in 2015, Dr. Amanda H. Goodson agreed to be my coach. At the time, there were so many gaps in what I needed to learn and do to grow. I used to dodge some of our conversations because Amanda was so open and so real, it pushed me to expose areas in my personal and professional life that I was not ready to deal with. However, there were shortcomings I knew I needed to face and address to become better at what I was doing and what I wanted to do.

Early in our engagement, Amanda asked me if I had a vision board. I didn't have one. Five years later, WeTECH Rocks was a reality and we were working on this book together with Odetta. In addition, I learned that I could focus so much on making sure I'm providing what others need to develop them as leaders that I sometimes forget my existing strategic placement as a leader of engineers and technical professionals.

If I had not had those experiences with Amanda as my coach, I would have certainly missed out on opportunities like this. Amanda set the bar higher than I ever expected or anticipated, and with her coaching, I got there! I'm now even serving on boards and working with presidents and CEOs of major

> organizations and companies because I took
> the time to invest in getting better and ready.

As a woman in technology, you need both coaching and mentoring, depending on your goals and strategies for your professional career. Mentorship deposits while coaching pulls out. Then there are also technical *sponsors* who go into a room and have your picture on their badge. They find opportunities to give you visibility and put your name out there to ensure you are successful. Sponsors are those who are on top of the mountain, see you coming up the slope, and say, "Here is a better way to come up," then shine a beacon and keep it flashing to show you which way to go. Using the rubber band example, a sponsor can show you how to navigate upward, relieving the tension and making that journey easier as they show others you have the capacity to move higher and stretch to different places. Any resistance is minimized, and your upward trajectory is smoother.

Have an attitude for altitude— and soar!

*E*lijah Cummings was a congressman and civil rights advocate who served in the United States House of Representatives for Maryland's seventh congressional district from 1996 until his death in October 2019. He once spoke to a group through Advancing Minorities' Interest in Engineering and shared an incredible story and message.

The son of sharecroppers, Elijah said he was he was labeled with a learning disability and placed in special education classes for six years as a child. He also liked to talk (a foreshadowing, surely, of his future as a politician) and was sent to the principal's office every now and then. He said one of his counselor's asked him. "Hey, Elijah? What do you want to be when you grow up?"

He said proudly, "I want to be a lawyer."

The counselor responded, "You can't be a lawyer. You're not smart enough. People like you are not lawyers."

One day, while sequestered to the library after a reprimand for talking too much, another teacher came and asked Elijah, "What are your dreams? What do you want to become?"

Elijah repeated, "I want to be a lawyer."

The teacher made a commitment to Elijah that, if he would study and complete his assignments, he would personally coach him. Elijah agreed, figuring it was better than going to the principal's office.

Elijah ultimately graduated from high school and college with honors and from law school at Howard University magna cum laude. What about the school counselor? One day after Elijah had graduated and was working for a law firm, the counselor needed a lawyer. Elijah was assigned to the case and won it. Afterward, as the counselor was shaking his hand, Elijah said, "You don't remember me, do you?"

"No. Who are you?"

"I was the young kid you told couldn't be a lawyer and couldn't make it. Thank you for teaching me not to give out, not to give in, and not to give up—because my attitude is everything."

Years later, the teacher who had coached Elijah was old and frail, and Elijah visited him in the hospital. The old man was so tired and sick he was trembling, but he was grateful Elijah was there. As he started to fade off to sleep, Elijah tiptoed to the door and was getting ready to leave when the teacher said, "Elijah, are you leaving?"

"Yes," he said, "but I have one last thing. I am so grateful for everything you did for me. Thank you so very much for being part of my destiny."

The old man shook his head. "You are thanking me, Elijah, for being part of your destiny? It is me that should thank *you* for being a part of mine."

———

Through WeTECH Rocks and in *TLC: Teams, Leaders, and Change*, we hope we have given you the direction and the tools to not give out, not give in, and not give up. Your attitude will make a difference, and it will give you the altitude to soar as a woman in technology. If you can see yourself as being impacted by someone else in a way that changes your destiny, then you can find things to be a part of, and identify people you can help, so that someday they will thank *you* for being a part of their destiny.

We want to take this opportunity to thank you for being part of our destiny by reading this book—and we are thrilled to be a part of your destiny. You don't have to do it alone, and you *will* not do it alone. We will be among those who will teach you and reach you so that you can excel and propel yourself forward—with the knowledge that you can do anything that you set your mind to do.

For much of her young life, Amanda struggled with self-confidence. In school, her teachers did not think she was as intelligent as the other kids, and she didn't perform well on standardized tests. Still, she made better than average grades in all her courses, even if her teachers didn't recognize it. As a high school student, Amanda took part in a math competition and won eighth place in the entire school, the only girl and African American in the top ten. Yet she was still shocked when her name was called because she never viewed herself as being smart. Everyone had told her she was average, and that seeped away at her confidence.

As her senior year neared an end, Amanda realized she needed to decide what she was going to do as an adult. With her father's encouragement, she went to the library to research what professions paid well that might fit with her two interests, math and music. It was clear. Engineers and accountants generally made money while musicians didn't. So, she thought, engineering it is. She made an appointment with a high school counselor.

"Girls are not engineers," she was told, point blank. "Maybe you should go into the military or be a nurse."

That was her first experience with gender bias—and Amanda didn't give in to it. It wasn't that she thought there was anything wrong with being a nurse or in the military. She just knew those weren't the paths for her. So, despite her counselor's advice to not go to a traditional four-year college, she decided she would, starting with Preface, a summer pre-engineering program for high school graduates at Tuskegee University. It was there where her instructors first told her she was smart, and it inspired her to apply herself and study like she never had before.

From that moment on, Amanda decided she was not going to be average and that she was going to excel as a young engineering student at Tuskegee—a decision that ultimately landed her with her first post-graduation job at NASA.

———

You can succeed! You can be great! With each chapter, we've strived to give you something you can use to grow and develop.

———

1. **Identify your style, exploit your strengths, and recognize your differences to manage them well.** You should have a great command of your leadership style as well as your strengths and differences. If you haven't

already done so, stop now and identify how you will exploit those to become a better leader. We know you can do it and get great at it! We look forward to hearing what you have learned from your successes.

———

2. **Acquire cultural awareness and clarity to become more effective as a technical professional.** In this day and age, cultural awareness and clarity is very important for a technical professional because people in the workforce have diverse ways they assess and implement goals. So, it is critical for you to understand your workplace network and culture. Pivot to navigate your own technical and professional awareness to become the best teammate and leader you can be. Don't forget that a SPOT analysis can help you exploit your strengths and recognize your differences. Also, be sure to leverage GRACE to mitigate conflict.

———

3. **Communicate and connect early and often by speaking up and speaking out in professional meetings and by using technical writing to create excellence and advance your career.** Communication

is how we verbally or nonverbally pass information from one person to another, and how well you communicate will affect your ability to lead. Evoke an "I belong here" mentality and behavior. You are an executive in the making (if you're not already there), so make sure to use every advantage to become a better overall communicator, writer, and a person of excellence.

———

4. **Strive to be great!** Remember, never give up. Never give out. Never give in. You have made a mark and an indelible impression on those around you. Don't take that for granted. Strive to be great, and continue on your successful journey. When something shows up that doesn't look like greatness, handle it right away, overcome it, and own the room! On your journey to be great, use passion, people, preparation, position, purpose, and proclivity to create a discipline that will take you to greatness.

———

5. **Use strategic planning, which is foundational and critical to map your future.** Now that you can see the big picture and your desired end state, leverage this as

an advantage for you and your capable team. Your big and bold ideas as a leader will cause you to reach your objectives in a meaningful way. Use every tool and resource available to you to make your competition irrelevant. Use your FRESH WILL to propel you forward.

———

6. **Invest in technical and professional development, knowing that continuous improvement is the key to success.** As you invest in yourself, there are classes that you may want to take or books that you want to read. Do not allow your skills to become stale and stagnant. Technical and professional knowledge is waiting for you to embrace it and use it to continuously improve. Relevant technology is one of your greatest keys to success as you AIM HIGHER to continue to invest in your technical and professional development.

———

7. **Receive ongoing technical coaching to give you best practices, and mentors to help you navigate to places you never thought you'd go.** We believe that you are intelligent and bright, and you will make great things happen for the people around you. As a

leader, you will effectively lead teams. As a team, you will be a part of an elite class of people who deliver outstanding value. So, receive coaching and mentoring when you need to keep that strategic edge. They will help you to become more knowledgeable and skilled. They will also make you an effective leader who adds incredible value to your organization.

———

Your destiny is waiting. Are you ready? We say you are! It has truly been our pleasure to come together for such a time as this to present this information in a way that is meaningful to you and that will cause your life to change.

Now go! Bring TLC to your domain—and soar!

Author biographies

Dr. Amanda H. Goodson is a ground-breaking aerospace engineer who soared to become the first woman to hold the position of Director of Safety and Mission Assurance out of the Marshall Space Flight Center at NASA. Transformed from a young African American girl who was told by her teacher that she would not amount to much, Dr. Goodson uses her unstoppable "can do" spirit to inspire others to achieve their goals regardless of the obstacles.

Noted nationally for her achievements, Dr.

Goodson is the recipient of the Southwest Alliance for Excellence Leadership Award; the Tucson's Woman on the Move Award for Leadership, Achievements, and Continuous Improvement in the Workplace; the Exceptional Service Medal; the Federal Employee Supervisor of the Year Award; and the Director's Commendation for Leadership Excellence in Safety, Quality, and Mission Assurance at NASA.

Dr. Goodson has served as the board of director's chair for Advancing Minorities' Interest in Engineering (AMIE), a national consortium for industry, government, and academia. In addition to serving in a leadership position for a Fortune 500 aerospace company, Dr. Goodson is the senior pastor at Trinity Temple CME Church in Tucson, Arizona.

Dr. Yvette Rice is a results-oriented executive with front end of the business acumen and demonstrated success in the development and execution of technical communications strategies that maximize team effectiveness, develop talent, and drive lasting cultural change. Dr. Rice uses her executive leadership and communication skills to direct and oversee technical writing teams that develop win-win strategies for the acquisition of contracts with city, state, and federal governments as well as for businesses and non-profits nationally and internationally. A published author, Dr. Rice utilizes her 30 years of expertise, influence, and business insight to partner with senior executives of large corporations to produce books, white papers, and prolific articles related to leadership development, mentorship, and coaching along with business advancement and developing and deploying platforms globally.

Dr. Rice majored in engineering via a scholarship program for Women in STEM through the University of Alabama in the 1970s. She uses her experiences to encourage other young women to consider a STEM career. Keeping up a family tradition of STEM majors, Dr. Rice's son, Samuel Christopher, and daughter, Sharné, are both engineers.

Odetta Scott is a certified Six Sigma Black Belt and serves as an Associate Director of Engineering Effectiveness for a Fortune 500 aerospace company. Scott attended the United States Naval Academy for two years before an illness ended her dreams of becoming an astronaut. Changing those aspirations, Scott chose to help others pursue their dreams of space flight by completing her bachelor's degree in mechanical engineering technology at Texas A&M University. Scott obtained a

Master of Business Administration degree from Jackson State University and earned a master's degree in organization development from Pepperdine University.

A valued professional development consultant, Scott spearheads a Leading Inspired Females in Technology recognition and retention sub-team. She also serves in organizations such as Advancing Minorities Interest in Engineering, the Society of Women Engineers, and as a STEM ambassador. An author, advisor, mentor, and lecturer, Scott fuels the development of individuals at all levels while driving transformational cultures in professional and business settings. Scott uses her "each one reach one" mentality to drive herself as a woman and as a leader to fulfill her God-given passion to better herself and others.

Capabilities + Book List

WHO WE ARE

WeTECH Rocks is a Women in Technology consortium with a vision of becoming a global leader in bridging the gap for the advancement of women in technology. Our mission includes accelerating value delivery, skills development, and strategy execution for leaders and entrepreneurs in Science, Technology, Engineering, and Mathematics (STEM). Our goal is to impart, in women and young women who are emerging in business, industry, the community, and in academia, how to become better leaders.

WHAT WE DO

As a catalyst for women in technical careers, we provide support in ways that build confidence, strengthening them in their respective fields by:

- Developing best practices and solutions for women to secure better jobs and achieve better placement in STEM in the areas they desire.
- Creating opportunities where senior women leaders in the community can come and share ideas and wisdom through mentorship and coaching.
- Facilitating a transfer of knowledge and skills to help young women develop greater competence and confidence.
- Rendering competition irrelevant because of the inimitable skills these women bring to the table.

CORE CAPABILITIES

Professional and leadership development
Effective verbal and written
communication development
Presentation skills
Conflict management
Leadership acumen
Personal branding
Emotional Intelligence
Self-awareness
Confidence
Critical thinking

SPECIALIZING IN THE FOLLOWING AREAS:

Leadership development for minorities
Six Sigma methodology expertise
Data capture and analysis
Effective communication skills development
Executive coaching
Mentoring
Strategic planning
Cultural awareness

If you are interested and want more information, you can reach WeTECH Rocks at wetechrocks@gmail.com.

Books written by or including content from your three TLC authors

Professional and Personal Development

Astronomical Leadership by Dr. Amanda H. Goodson

The Purpose and Power of Mentorship, compiled by Dr. Amanda H. Goodson and Marvin Carolina, Jr.

12 Power Principles for Administrative Professionals by Dr. Amanda H. Goodson

Soar to Your Destiny: 5 Winning Success Strategies by Dr. Amanda H. Goodson

How to Unlock Your Full Potential: 11 Keys to Leader Success by Dr. Amanda H. Goodson, Odetta Scott, and Frederick Cross

Inspirational

Women in Leadership: Living Beyond Challenges, compiled by Dr. Amanda H. Goodson Amanda and Dr. Yvette Rice

Phenomenal Prayer: Activating Who You Are by Dr. Amanda H. Goodson

Switch to Holiness Workbook: 12 Actions to be Your Best by Dr. Amanda H. Goodson

Spiritual Intelligence by Dr. Amanda H. Goodson

The Power Behind Your Need is in Your Seed by Dr. Amanda H. Goodson

Financial Healing from the Inside Out by Angela C. Preston and Dr. Amanda H. Goodson

Chosen to Worship: 21 Days of Prayer by Dr. Amanda H. Goodson

Powerful People Follow Christ by Dr. Amanda H. Goodson

Mountain Moving Made Easy by Dr. Yvette Rice

Mountain Moving Made Easy Workbook by Dr. Yvette Rice

www.ingramcontent.com/pod-product-compliance
Lightning Source LLC
Chambersburg PA
CBHW060616200326
41521CB00007B/783